DIY BIOGAS

Make and Use Your Own Renewable Natural Gas

By Paul Scheckel

First published by Paul Scheckel, 2022
Parsec Energy Consulting
Copyright © 2022 by Paul Scheckel

ISBN: 978-1-7362902-3-1

This book is for entertainment purposes only. It is not intended to be a comprehensive text on all the intricacies involved in successfully managing the information contained herein. Adequate skills and knowledge in multiple disciplines are required in order to safely approach the subject matter presented. If you do not have the appropriate level of knowledge and skills, seek training, or enlist a competent professional.

The author cannot be held liable or accountable for how the information in this book is used. Following the advice in this book can lead to fire, personal injury, and damage to property. You alone are responsible for how you use the information presented, and you assume any and all liability for damage to people or equipment that may result. It is your responsibility to determine the suitability of any project, parts, assembly, and any and all results or outcomes, to be used for any purpose whether presented in this book or not. It is up to you to use tools and equipment properly and to take proper precautions with chemicals, mechanical equipment, electrical service, combustible gas handling, associated materials, and procedures. You alone are responsible for injuries to yourself or others or for damage to equipment or property.

Have fun, be safe!

Table of Contents

Introduction: What is Biogas?

Biogas is a mixture of gases formed anywhere organic material decomposes in the absence of oxygen. Underwater, deep in a landfill, bubbling out of municipal solid waste, or in the guts of animals (including you), biogas (aka swamp gas) is produced through the biological and chemical processes of anaerobic digestion (AD). All that is a long-winded way to describe a fart. Yes, your farts – and your dog's – are biogas, and they will burn. How much flame is produced depends on what you (or your dog) eat and how efficient your digestive system is at dealing with your diet. This decomposition of organic matter happens without any outside help, but we can assist nature by providing the ideal environment to maximize gas production.

Anaerobic digestion is the decomposition (digestion) of carbohydrates in an oxygen-free (anaerobic) environment. It begins with a process similar to the fermentation of alcohol, but without oxygen AD continues past fermentation. In fact, oxygen is toxic to the process in that it inhibits the growth of methane-producing microbes, known as **methanogens**, which are ultimately what we want to encourage in order to produce combustible biogas.

The main ingredient of biogas produced in this controlled environment is methane. Methane is a hydrocarbon made up of one molecule of carbon and four molecules of hydrogen (CH_4). Methane is the primary component of natural gas commonly used for cooking and heating, although biogas is not as energy dense as natural gas. The methane content of the biogas you

1

make will probably range from 50 to 80 percent, compared to about 70 to over 90 percent with utility-supplied natural gas. Natural gas also contains other combustible gases such as butane, propane, and ethane, while biogas does not.

The exact makeup of biogas depends in part on what you feed the bio-digester and in turn, what was fed to the producers of those ingredients. Non-combustible impurities of biogas are primarily carbon dioxide (CO_2), water vapor, and possibly trace amounts of hydrogen sulphide (H_2S). If air leaks into the system and contaminates the process, nitrogen will further dilute the biogas. Other impurities may be formed as well, depending on your feedstock and the efficiency of the process. These impurities can be removed if desired, depending on how you intend to use the biogas. Note that the use of the term "biogas" refers to the gas produced by your biogas generator, and the term "methane" refers specifically to CH_4. They are not the same. If you've ever lit a fart (and you should), it's only the methane that burns.

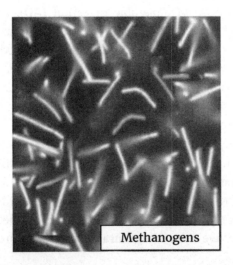

Methanogens

Note that throughout this book you'll notice that some text appears as a hyperlink. The Kindle version of this book contains hot-linked URLs for some products and information.

The Basics of Producing Biogas

A good way to conceptualize the making of biogas is to compare it to the digestive process of animals. When animals eat, food is broken down through mixing with saliva (water and enzymes), then transferred to an anaerobic environment (the gut) where it's digested. The results are energy made available for the animal, plus liquid, solid, and gaseous waste. For our purposes, we want to control the process and the environment to maximize the "waste" gas production and its energy potential.

To produce biogas, organic material (feedstock) such as animal manure or vegetation is mixed with water and inoculated with a starting culture. Inoculation can be compared to pitching yeast when brewing beer or baking bread in order to direct the fermentation results. Everything is then sealed in an airtight container. Maintain a temperature within the container that is close to the temperature inside an animal (around 100°F) and in about a week gas will start to flow. But it's not quite combustible gas yet.

The airtight container where this process is captured and controlled is called an "anaerobic digester" or "biogas generator". I tend to use the term 'generator' to refer to the entire system that assumes intention and design toward producing something useful. The digester vessel is the part of the generator where the ideal conditions are maintained to encourage anaerobic digestion of organic material, a process that can happen even without our intention or intervention under the right conditions.

While design specifics can vary, the main parts of a methane generator include the air-tight digester vessel to hold the material while it decomposes, an inlet for feeding, an effluent outlet to remove digested solids and liquids (digestate), a gas outlet, and a way to store and control delivery of the gas. You can make a biogas in a plastic bottle, a 5-gallon bucket, a 55-gallon barrel, or a 10,000 square foot manure handling facility; the concept is the same. However, any digester vessel less than 200 gallons should be considered experimental in that it will not produce enough biogas to be useful for any practical purpose other than testing recipes and perhaps occasional gas grilling. I highly recommend starting small so that you can get a feel for the process with minimal effort and expense.

Challenges

Be warned! Having a biogas generator is like having another mouth to feed – it takes time and attention. You must provide the right environment along with the correct type and quantity of food to keep the microbes alive and healthy. Other practical challenges are the availability of, and ability to collect, handle, transport, and process enough locally available organic material to make a useful amount of gas. You'll need a way to store the gas, provide a mechanism to control the gas pressure and flow, and plumb it all together to safely deliver the correct amount of gas at the right pressure to the burner. Depending on the end-use of the gas you produce, you may need to modify the burner so that it can handle a less energy-dense gas that contains a higher level of impurities. If the intended use is to fuel an engine, you'll want to purify the gas so that it's a high percentage of methane by removing diluting and potentially damaging components of the gas.

Keep it Simple

The biological and chemical processes of AD, along with all the nuances of feedstock variables, are complex. However, if you dwell on the complexity of the science, you may never get started. Save that step for

4

when you turn pro. Anaerobic digestion is a natural process of decay that wants to happen by itself — any encouragement you offer can only be helpful. If you're anxious to get something going now, find a container of any size that can be sealed tightly, fill it 50/50 with water and any sort of organic material you can find (leaving about a third of the container unfilled), and put a home-brewer's airlock on top. Keep it warm, and you'll have some success in making biogas within a week or two. It may not contain much methane, but something will certainly happen!

How it works

Once your generator is filled with organic material and water, biochemical processes begin to happen. First the ingredients will breakdown and ferment, then acids will begin to form, followed by the desired methane production. There are four stages in the breakdown of organic material within the slurry of your biogas generator. These four stages can be separated into two phases: acid formation and methane formation. The waste of one stage feeds the next without any outside intervention. Once a digester is operating and producing gas, these processes happen simultaneously inside the soup of organic material rather than as discrete sets of chemical reactions requiring different environments or treatments.

Hydrolysis starts when water is mixed with organic material. Hydrolysis is the enzymatic breakdown of complex proteins, carbohydrates, fats and oils into amino acids, simple sugars, and fatty acids. The broken down (depolymerized) material is now in a much more chemically accessible form and ready to be fermented by acid forming bacteria.

Acidogenesis, or fermentation, happens when acid forming bacteria oxidize the simple compounds formed during hydrolysis to create carbon dioxide, hydrogen, ammonia, and organic acids.

Acetogenesis is the conversion of organic acids into acetic acid. Acetic acid is the main ingredient in vinegar and is the food for the final stage of decomposition within the generator. Acid forming bacteria are fast breading and hearty, producing lots of CO_2.

Methanogenesis is the reproduction of methane producing microbes, or methanogens (single-celled, non-bacterial microorganisms from the

5

biological group *Archaea*). Methanogens use the hydrogen and CO_2 produced during the acid forming phases to create methane. In contrast to the acid formers, methanogens are slow to reproduce and extremely sensitive to temperature, pH, and the presence of oxygen. Importantly, methanogens need to be present in the initial feedstock loading of your digester.

Production Rate

There are many variables in all processes of generating biogas. Of primary importance is the recipe formation including the type and quality of feedstock that goes into it, and how well the internal environment is maintained. I point this out because it is almost impossible to calculate exactly how much gas will be produced from any given recipe. However, there are rules of thumb that offer enough guidance to point us in a generally useful direction. Once you get started, recipes and feeding can be adjusted based on your actual results.

Rule of thumb for biogas production: A well-managed generator may produce approximately its own volume of biogas each day. In terms of energy production, a bit of math is required.

- A 55-gallon drum has a volume of about 7.35 cubic feet.
- One cubic foot of pure methane contains 1,000 BTUs of energy. One BTU, or **British Thermal Unit,** is the amount of energy required to increase the temperature of one pound of water by 1 degree F. This is approximately the amount of energy released by burning one wooden kitchen match.
- Biogas containing 60% methane offers 600 BTUs of energy for each cubic foot. The remaining 40% of noncombustible gasses serve only to dilute the energetic methane content.
- 7.35 cubic feet x 600 BTUs per cubic foot = 4,410 BTUs of gas energy produced in one day.

A typical gas cook stove might burn 15,000 BTUs of fuel in an hour on maximum heat. 15,000 BTUs per hour divided by 4,410 BTUs in storage means all the gas will burn in about 18 minutes. At that rate, your 55-gallon methane generator has produced enough gas to boil about 2 gallons of water (assuming 60% transfer efficiency between the energy in the flame and the water in the pot).

This might be enough in some cases, but in a practical sense a small family with modest daily cooking needs will require a warm, well-fed, 200-gallon (27 cubic foot) methane generator. This much biogas represents about 16,000 BTUs and offers about an hour of cook time, or enough energy to boil around 8 gallons of water.

Variables

The quantity and quality of the biogas you make depends on the nutrient value of the feedstock and how well the microbes convert the available nutrients into methane. For practical purposes, biogas production and quality are functions of your specific recipe and generator management. Important things to understand about making biogas and maximizing production are:

- Recipe development and the carbon-to-nitrogen ratio of ingredients
- Water content of ingredients
- Solids, liquids, and digestible quality
- Temperature
- Feeding rate
- Retention time
- pH
- Mixing

We'll cover all of these topics, but as you can see, any estimate of methane yield for each unit of digestible material has quite a few variables, all of which can be adjusted and controlled. There are published lists indicating specific values for some of the above variables (as in Appendix A), but they are just that – estimates. Consider them to be guidelines from which you can begin to develop your recipe. Your actual production *will* vary. I encourage you to not get lost in the numbers or details and simply experiment. You can learn at least as much by doing as by studying!

Biogas Recipe

M any combinations of organic materials can be digested including vegetables, food scraps, grass clippings, animal manure, meat, slaughterhouse waste, and fats – almost anything that contains carbon and/or nitrogen. Any recipe must achieve a suitable balance between carbon and nitrogen. Avoid using too much woody products, like wood chips and straw, containing large amounts of lignin (which is resistant to microbial breakdown and tends to clog up the digestion process). Also avoid material that may be contaminated with heavy metals or other toxins, and materials with large amounts of ammonia or sulphur. The ideal ingredients are those materials you have a plentiful, convenient, and consistent supply of, so that you can make consistent and useful quantities of biogas while maintaining a healthy biome in your digester. If you have experience with mixing compost (aerobic decomposition of organic materials), you already have a good idea of what the recipe needs to be: if you can compost it, you can digest it, the only difference is the absence or presence of oxygen.

The Carbon to Nitrogen (C:N) Ratio

Organic material can be classified by how much carbon and nitrogen are in its makeup, and the ratio between those two elements. Carbon is a source of energy for microbes, while nitrogen is needed for protein to build cell structure. The content ratio that compares them is expressed as C:N. The conventional way to represent this is to compare the carbon content against one part of nitrogen, so that the N value will always be 1. You may find data indicating the percentage of carbon and nitrogen in a material. In that case, calculate the C:N ratio by dividing the carbon percentage by the nitrogen percentage. For example: chicken manure contains (on average) 37.8% carbon and 2.7% nitrogen. The C:N ratio is 37.8/2.7 = 14, or 14:1, meaning that there is 14 times more carbon than nitrogen.

For anaerobic digestion, a C:N ratio between 20:1 and 30:1 is suitable, with the higher C:N range being ideal. Too much carbon slows the decomposition, while too little means organisms don't grow. Too much nitrogen produces ammonia and too little means that not enough of the right kind of microorganisms will grow. There are exceptions though, for example animal manures are often good methane producers even though the C:N ratio isn't ideal. Don't worry too much about getting the C:N ratio perfect at first, since a wide range of C:N will produce useful results. Experiment and find a recipe that works well with available ingredients, then perfect the recipe and processing for maximum gas production.

A visual way to think about the carbon nitrogen ratio is in terms of brown and green. In general, things that are brown are high in carbon. This includes cardboard, wood chips, corn stalks, dry leaves, pine needles, and straw. Green things are generally higher in nitrogen, including food and garden wastes, grass, seaweed, and manure (an obvious exception to the general color rule).

The table in Appendix A, *Characteristics of Raw Materials* (CRM), lists approximate average C:N ratios, moisture content, weight, and estimated *volatile solids* (more on this term soon) content of common organic materials. Any combination of these can be mixed to provide the ideal ratio.

Keep in mind that the C:N ratio describes just one aspect of the chemical makeup of the material – it does not mean that you need 30 times more brown material than green. The actual amounts of carbon and nitrogen in any material will vary depending upon the specific makeup and age of that material, along with everything that went into its production such as soil conditions, nutrients available to that crop, and (in the case of manures) the animal's diet. The data provided should be taken only as a general approximation of C:N ratios. Note that the terms "VS" and "TS" shown in the table are discussed in the section on Solids and Liquids.

Calculating C:N Ratio in a Recipe

You can determine the C:N ratio of any combination of ingredients, and thus the relative quantity of each ingredient required in your recipe, if you know the C:N ratio and the approximate moisture content of each material.

Note that C:N ratios in the CRM table are given in dry weight, not wet weight, and not by volume. This is because moisture content varies greatly, even within samples of the same material. You can only get away with using volume measurements if the water content of the ingredients is the same. Factoring in moisture content means an extra step in the analysis, and it will be an estimate, but it's a simple step when weighing ingredients to add to your recipe.

To calculate the C:N ratio of your recipe, make a list of each of your available ingredients and refer to the CRM table. Then, for each ingredient:

1. Look up the carbon (C) value (the C value is the same as the C:N ratio because the N value is always 1).
2. Determine the wet weight.
3. Look up the moisture content percentage.
4. Figure out the dry weight of each ingredient using this formula:

    Wet weight x (1 – moisture content percentage) = dry weight

5. Multiply the carbon value of the ingredient by its dry weight to find the carbon "units". The units reflect the weight adjusted percentage of carbon.
6. Add up all the dry weights (from Step 4) of the ingredients.
7. Add up all the carbon units of the ingredients (from step 5).
8. Divide the total carbon units (from step 7) by the total dry weight (from step 6). The number you get is the carbon value (C) of your recipe's C:N ratio, where the nitrogen value is always 1.

Recipe Example

Here's an example recipe using materials you might have around your homestead. The quantities are relative and can be scaled up or down as needed. The weight system you use doesn't matter – it could just as easily be kilograms, ounces, tons, or drams – just be consistent. We'll develop a recipe using the values in the CRM table for chicken droppings, food waste, and a small amount of carbon-rich paper product. Weight is in pounds (lbs). Don't worry too much about precision as it is not obtainable, or required, outside of a lab.

The table below shows all the values we'll need to analyse our recipe.

| | | | | | Carbon Units |
| | Carbon | Wet | Moisture | | Carbon x |
Ingredient	Value	Weight	Content	Dry Weight	dry weight
broiler litter	14	1.0	37%	0.63	8.82
kitchen scraps	15	1.0	69%	0.31	4.65
wood/paper	625	0.020	5.5%	0.0189	11.81

Table 1: Recipe Evaluation C:N ratio

Total wet weight	2.02
Total dry weight	0.96
Total carbon units	25.3
C:N ratio of recipe	26.4

Broiler litter is chicken droppings with some bedding material mixed in. On average, it has a C:N ratio of 14:1, a moisture content of 37%, and we'll estimate 1 pound (wet) to add to your recipe. Pure chicken manure has a higher moisture content (closer to 70%) and lower C:N ratio (approx. 6%).

1 lb wet weight x $(1 - 0.37)$ = .63 lbs dry weight

Next, multiply the carbon value in the C:N ratio by the dry weight to find the carbon units:

14 x .63 = 8.82

Food waste/kitchen scraps: 1 lb wet; C:N = 15; 69% moisture; 0.31 lbs dry; 4.65 carbon units

Newsprint: .02 lb; C:N = 625; 5.5% moisture; .019 lb dry; 11.81 carbon units

Total wet weight of all ingredients = 2.02 lb

Total dry weight of all ingredients = 0.96 lb

Carbon units total = 25.3

25.3 carbon units ÷ 0.96 lb dry weight = 26.4

This means the C:N ratio is 26.4:1, right where we want it.

Consider this example a place to start and not the final word on the available carbon and nitrogen in any given recipe. Experiment to find out what works best for you using your specific ingredients. Notice the very small amount of newspaper required due to its high carbon content (the weight of about 1 paper towel was added to the example above). Small changes in the addition of woody material will make large differences in the C:N ratio. Again, be aware that not all material digests equally well, even if the C:N ratio is identical, as it's only one part of the chemistry. Some materials are more biologically available than others.

Solids and Liquids

Most materials contain water, many contain mostly water. Water is an aid to digestion, but it can not be digested. Organic materials might be 75% water by weight, more or less, and if you were to evaporate all the water from a material, you'd be left with 25% **total solids** (TS). Only a portion of the total solids in organic material are available to assist in the bacterial breakdown processes of AD. These are called **volatile solids**.

Volatile solids (VS) are the portion of total solids available for anaerobic digestion. VS is the important piece of the feedstock equation in determining how much gas can potentially be produced for any given material.

Fixed solids (FS) If you evaporated all the water from a material, and then burned the remaining total solids in a fire, you would have a pile of ashes called **fixed solids** that are unavailable as food for the digesting microbes. Everything that burned is VS, and that might be 80%, more or less, of your TS.

Chemically speaking, this approach is a bit rough in determining exactly how much of which part of the material gets digested, but it's the best we can do at this point without a lab. If you take this process to a higher level of professionalism and require analysis to maximize your yield, search for agricultural testing services near you and have them test your recipe sample.

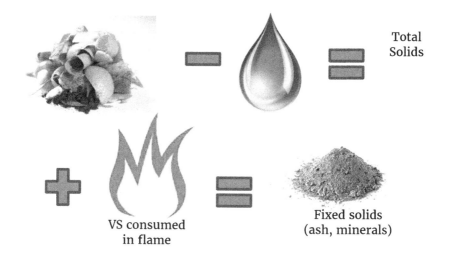

Total
Solids

VS consumed
in flame

Fixed solids
(ash, minerals)

VS and Gas Production

Available VS in your digester slurry is directly related to gas production. One pound of volatile solids can theoretically yield a maximum of 30.5 cubic feet of biogas under ideal conditions. In reality, anywhere from 10 to 60% of the VS will be converted in the digesting process, so the practical result is that you can expect anywhere between about 3 and 18 cubic feet of biogas production per pound of VS. The gas will likely be somewhere between 55 and 80% methane, so the methane yield for each pound of VS consumed in the generator will range from 1.7 to 14 cubic feet. That's a range of 1,700 to 14,000 BTUs of energy.

Studies of biogas yields from various feedstocks are generally represented in terms of cubic feet of methane produced for each pound of VS converted under the specific conditions of that test. Expect that your results will be different from anything you see published. The specific composition and availability of VS within organic wastes, along with the environment within the digester and how you manage the whole process will all affect gas production.

Digester Care and Feeding

Now that you've developed a recipe and understand what the variables are, you can begin to feed the digester. Solid material will decompose better when cut, ground, chopped, or shredded into one-inch (or smaller) bits. More surface area available to the microbes allows for better digestion of organic material, yielding more efficient gas production. A kitchen sink food grinder, old food processor, or a yard waste chipper/mulcher can be convenient ways to prep large amounts of material. Fibrous material may digest more readily if it has been allowed to age for a few days before putting it into the digester. This allows for fungal and bacterial breakdown of the fibre, but don't age for too long or energy will be lost.

Add enough water to make a pourable slurry and add it into the digester. The general rule of thumb is to add the same amount of water as solid material, but this ratio will vary according to how much water is already in the organic material in your recipe, and how much fibre is in the organic matter. More fibre requires more water to break it down, but less water means more room for digestible solids, and thus more gas.

The commonly acceptable range of total solids for optimum biogas production is between 2 and 10 per cent, meaning that 90 to 98% of the material inside your digester, including the moisture in the material itself, can be water. Other considerations around how much water to add are material handling and digester volume.

Inoculation

When you first load the digester, you will need to inoculate it with a culture of methane producing organisms (methanogens). These microbes occur naturally in many animal manures, so if you're using manure you don't need to worry about it. But if you want to digest only food scraps or grass clippings, you'll need to inoculate it initially to get the biological processes going. A good culture can come from fresh farm animal manure (ideally cow or pig manure), slurry from another operating digester, a shovel full from the bottom of a compost pile that has not been turned in a long time, or the muck from the bottom of a biologically active swamp. Once added, the methanogens will reproduce, and you can stop feeding manure after the initial start-up. Methanogens die quickly when exposed to oxygen, so the manure used as the inoculant needs to be very fresh! I've successfully re-started a dormant digester after a couple of months without feeding, so the methanogens will survive on small amounts of material so long as no air is allowed to enter the system.

Starting Up

Fill the digester to about 80% full of the slurry of water and organic matter from your recipe. More air space is okay if you're starting with just a small batch of material, but it will take a bit longer for your gas quality to increase so that it will burn. At first, the generator will produce a large amount of CO_2 as the  available oxygen is consumed and the methane producers catch up to the CO_2 and acid producers. This means that the first batch of gas will be mostly CO_2 and will not burn. Once the oxygen in the generator is consumed by chemistry, the process stabilizes and shifts from *aerobic* (microbial breakdown *with* oxygen, as in composting) to *anaerobic* digestion.

Temperature

Temperature management is a critical detail in the generation of biogas. Different groups of methanogens have been identified that respond to different temperature regimes. You must choose which regime you wish to work with and design your system and processes accordingly:

Psychrophilic methanogens survive at cooler temperatures, above 32°F

Mesophilic methanogens are active at temperatures between 70 and 105°F

Thermophilic methanogens will dominate at temperatures between 105 and 140°F

In the cooler psychrophilic range, gas production rate will be quite low. Mesophilic methanogens are much more tolerant than thermophilic, and faster producers than psychrophilic. Mesophilic methanogens may produce gas down to 70°F, but at a very slow rate. One advantage to the thermophilic methane producers is that they are much faster at digesting, requiring only about half the material retention time to achieve similar gas production compared to the mesophilic range, thus requiring a smaller volume digesting vessel. To achieve this efficiency however, there is a penalty in energy consumed to provide the required heat. For our purposes we'll focus on the more easily managed and better understood mesophilic temperature range.

The mesophilic conditions to maintain within the digester are similar to those inside an animal's gut; that is to say, oxygen-free with a temperature of around 98°F (37°C), plus or minus a few degrees, for maximum gas production. Biological activity within the digester will generate some heat but depending on your climate you may need to supply heat to the digester.

To reduce the amount of external heat required, place the system in the sun or inside a greenhouse, insulate the exterior, and/or cover it with black material to absorb solar heat. To produce gas during the winter in cold climates you'll need to provide a source of heat to keep the slurry warm. A large biogas generator may produce enough gas so that some of it can be used to heat water which can be circulated inside the digester. Submersible electric water heaters designed to safely heat plastic animal waterers can be used, or you can build a separate solar water heater with an internal or external heat exchanger. Find simple plans for that project in the Homeowner's Energy Handbook (Storey Publishing).

You'll need to weigh the costs of providing heat against the benefits of gas production rate. Alternatively, if you live in a hot climate you may want to provide some shade so that the temperature inside the digester stays below 110°F.

Retention Time

In most cases, material you put into a well-maintained methane generator operating in the mesophilic range will be well digested in about a month. You might get more gas from the same material over a longer period, but the production rate will fall off over time.

Handling more material for a longer period requires a larger digester. Retention time (also called *hydraulic retention time* or HRT) means how long the material stays inside the digester vessel where it's broken down over time. Retention time is determined by how quickly the material breaks down, the gas output requirement, and the volume of your digester. HRT and digester volume will determine the rate at which you feed the digester, so that material is retained for the optimum amount of time after having delivered an optimum amount of gas.

For example, if you have a 55-gallon drum (with a volume of 7.35 cubic feet) and you need a retention time of 30 days to produce the optimum amount of gas from a specific recipe, you want to move 55 gallons of material through it over a period of 30 days, for a feeding rate of just under 2 gallons (.27 cubic feet) per day. Every time you add 2 gallons, 2 gallons will flow out the effluent tube. The gas production is limited by the digestible content (VS) of the organic matter you add. Optimum retention time will vary with recipe and temperature, so experiment and find out what works best in your situation. For maximum efficiency you want to match retention time with gas production, which in turn determines the rate at which you feed. As you work with your digester and materials, it will be fairly easy to see the gas production rate as your storage vessel gains gas volume. Feeding rate is related to *loading rate*.

Loading Rate

Loading rate refers to the quantity of volatile solids added to the digester with each feeding, as well as the frequency of those feedings, and is closely linked with retention time. Loading rate is generally expressed in terms of pounds of VS per cubic foot of digester volume and is managed according to the properties of the material being used. If the diet you feed your digester is fairly consistent, you don't need to go through the exercise of calculating VS every time. After a while, you'll get a feel for how much and how often to feed, based on your recipe and observed results of gas production.

The loading rate of a digester vessel is a function of its volume and how well the methane forming microbes keep up with the acid forming bacteria. The rate will vary somewhat depending upon the digestibility of the feedstock(s) you're using, since not all VS behave in the same way. Assume that only about one-half of the VS added will be digested, and expect that amount to vary for different feedstocks (and even among the same feedstock), depending on its exact makeup. Different amounts of VS and gas can be expected from the same manure if that animal is on pasture for the summer and grain and hay for the winter.

Where to start

As a general rule, start the generator on a small amount of VS – say 0.1-pound VS for every cubic foot of generator capacity. As the methanogens become established, you can add more – perhaps up to .25 pounds of VS for each cubic foot of generator capacity. Some sources of VS will convert to gas at a higher rate and more completely than others, and you'll need to adjust the loading rate for best results. If gas production is fast and efficient, you can decrease retention time and move material through more quickly by increasing your loading rate. You can assess production over time by watching your gas storage vessel increase in size or displacement, or by burning it in the same way each day and recording the burn times.

Loading Rate and pH

Methanogenesis is a limiting factor in how much you feed your digester. Feeding too much allows the faster acid forming process to overtake the slower methane forming process. This leads to a low (acidic) pH level that essentially poisons the process by inhibiting methane forming activity and will ultimately shut down gas production. Check for this imbalance by testing the pH of the slurry, which should normally be right around 7 or 7.5. The test is easily done using pH paper or an electronic tester dipped into a sample of liquid digestate taken from the outflow tube. If the pH is below 6.5, the digester has gone sour. Stop feeding for a few days so that the methanogens can catch up, and test pH again. If things are really bad, you may need to add some baking soda to help bring the pH back up. Add small amounts, only ¼ to ½ cup at a time for a 55-gallon barrel digester, wait a day, and check pH again. When pH has returned to normal, gas should start bubbling out again. Start feeding and adjust the recipe and/or loading rate while monitoring pH as you make corrections.

Avoid indigestion

The ecosystem within the generator evolves to suit the recipe. If you change the diet in a sudden and dramatic way, the chemistry will react and ultimately readjust. Think of how you feel after trying a new food or travel to a foreign country and indulge in exotic local cuisine. Make small, incremental changes to your diet to avoid a belly ache.

Mixing

Some material fed to your digester will form a crusty layer of scum on the surface, reducing or preventing gas production. Daily mixing of the slurry is recommended. You can mix by shaking, stirring, or otherwise agitating the slurry by whatever practical means are available, as long as no air is introduced into the system. A power drill with a paint mixer attachment works well for mixing some slurries, and simply rocking a 55-gallon drum back and forth every day is effective. More agitation means less scum formation, and better mixing means more efficient gas production. Scum can be kept to a minimum by avoiding high amounts of lignin, and by chopping material into small pieces.

So, how much gas can you make?

Now that you understand the variables and nuances of producing biogas, we can go a little further in estimating gas production. Refer again to the CRM in Appendix A and look at the column "Average %VS of TS". We'll put values on that column and expand on the recipe we've already developed to calculate VS in that recipe. Remember, VS is directly related to gas production.

Of the 2.02 pounds (total) weight of the recipe, only 0.78 Lbs is VS. If you put that material into a 55-gallon drum (7.35 cubic feet) that is 80% full (5.9 cubic feet) of water, the loading rate works out to:

.78 ÷ 5.9 = .13 lb of VS per cubic foot of available digester volume. This is a good starting load rate.

					Carbon Units		
Ingredient	Carbon Value	Wet Weight	Moisture Content	Dry Weight (lb.)	Carbon x dry weight	% of TS that are VS	VS weight (lb.)
broiler litter	14	1.0	37%	0.63	8.82	77%	0.49
Food scraps	15	1.0	69%	0.31	4.65	90%	0.28
wood/paper	625	0.020	5.5%	0.0189	11.81	97%	0.02

Table 2: Recipe Evaluation VS and Gas Production

Assuming a 50-percent conversion of VS to biogas, gas produced by .78 lb VS over time works out like this:

(0.78 lbs VS) x (30.5 cubic feet of biogas per lb of VS) x (50% conversion rate) = 11.9 cubic feet of biogas.

Total wet weight	2.02
Total dry weight	0.96
total VS (lb.)	0.78
VS/CF (80% H2O)	0.133
VS to gas conversion rate	50%
theoretical max (CF/lb. VS)	30.5
CF biogas produced	11.9

If half the biogas is methane, energy production works out to just under 6,000 BTUs.

These are very general estimates and actual gas production may be quite a bit less. HRT, feeding rate, pH, and gas production will be checked and controlled based on your direct observations. Once the digester is loaded and operating, keep track of the rate of gas production by observing the daily displacement of the gas collection or storage vessel. When the production rate starts to decline, it's time to feed. Observe, measure, adjust as required.

Ok, but how much gas can you make, really?

Individual results are highly specific, but following are actual measurements from my 55-gallon hybrid system. It was actively producing gas for a while, then I stopped feeding until it stopped producing gas (about 2-weeks). Next, I ground up 6-pounds of mixed food scraps and added water to make about 3-gallons of slurry. The table below shows the daily production rate over 5 days.

day	Daily Cu Ft	biogas produced from food scraps
0	0	loaded with 6 lbs. food scraps in slurry
2	0.8	Maintain 100F for 2 days
3	0.9	Burned off but some sputtering
4	1.6	Production kicked in fast! Lost some gas.
5	0.8	Production starting to fall off
Total	4.1	**Cubic Feet**
	2,460	approximate total energy, Btus

Six pounds of food scraps translates into 1.08 pounds of VS assuming 80% moisture content. This is a fairly average loading rate of 0.18 lb VS/CF of digester volume. Theoretically, 1.08 lb of VS would yield about 33 cubic feet of gas under perfect laboratory conditions. Overall, the recorded results represent a conversion rate of 12.4% VS to gas after 5 days. If the production was measured for a full month, perhaps twice the amount recorded might have been produced, which would double the conversion rate to nearly 25% of theoretical maximum. Better results can be obtained with more consistent resource availability for feeding, tighter control of C:N ratio in the recipe, and by using pH as feedback to govern loading rate.

Types of Digesters

There are two general approaches to methane generator construction. The simplest design is a **batch processor**, but there are numerous advantages to a **continuous flow** process.

Batch

A batch digester breaks down one bucket full of material at a time. Material is loaded, the digester sealed, and gas is produced after several days at a rate and duration that depends on the recipe and the internal conditions. Gas is collected and used until the production drops below a usable level. When gas flow stops, the digester is emptied, effluent is removed and composted, and the process is repeated.

This relatively simple and inexpensive design is perfect for small test batches using easily managed volumes of material. However, more intensive material handling is required. Gas production can last anywhere from a week to a month, depending on the size, feedstock, internal environment, etc. You can build a simple batch processor using a tightly sealed bucket or barrel with a gas outlet port, much like a fermenting bucket used for beer or wine.

Continuous Flow

For managing a steady stream of digestible material offering a steady supply of gas production, the continuous flow process is the more practical approach. Continuous flow digesters are sometimes called *plug flow* because a "plug" of material is periodically fed to the digester, and another plug is simultaneously removed.

A continuous flow system has a feeding inlet on one end, an effluent outlet on the other, and a gas outlet pipe in between. Material is loaded at a certain rate and pushed through the digester over a period of time. It is sized according to the daily input volume and desired retention time. When new material is loaded in, the slurry flows through until it eventually reaches the outlet where the spent digestate is removed. Gas is withdrawn and used or stored as it is produced in a continuous process.

The digesting vessel could be an old piece of plastic culvert, a large concrete serpentine trough on a farm where manure is constantly loaded and pushed through, or a UV resistant polyethylene plastic bag (such as those used for bagging hay) utilizing PVC piping for inlet, outlet, and gas connection.

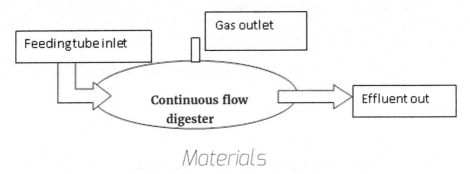

Materials

Given the potentially corrosive nature of biogas containing even small amounts of hydrogen sulfide (which may be produced when some manures are digested), avoid using materials that are susceptible to corrosion. Plastic barrels and PVC pipes will be more durable over time than metal barrels and galvanized pipe. Biogas contains water vapor, so slope gas pipes downward to allow the condensate to drain so as not to block gas flow.

What to do with the effluent

Effluent is the digested waste material from your biogas digester. It's a low odor blend of compostable solids and nutrient rich liquid. What you do with the effluent depends upon what went into it, how well the solids are digested, and perhaps where you live. You can apply the effluent directly to your garden or fields as a soil amendment, or you can compost it and further refine it for improving your soils. Some of the carbon has been consumed during the digesting process so the effluent doesn't have quite as much energy potential as the original feedstock, but nitrogen is not consumed.

One of the biggest concerns is the presence of pathogens in animal waste. Some pathogens will be destroyed in the mesophilic temperature range over a typical retention period, but some require much hotter temperatures to be completely destroyed. High-temperature composting (greater than 140°) generally eliminates most pathogens. If you're concerned, the only way to be sure is to have a sample tested at a reputable lab.

Using Biogas

Biogas can be used in place of natural gas, liquid propane gas (LPG) for space and water heating, lighting, cooking, to provide power by burning in an engine, or to fuel an absorption cooling system such as a gas refrigerator or chiller. Purified (scrubbed) biogas behaves just like natural gas, but if you choose not to scrub the gas, you'll need to modify the burners and delivery system to accept a reduced BTU gas containing CO_2 and water vapor. You can choose to modify either the gas or your equipment. Ideal combustion requires a burner that is designed specifically for a particular gas delivered at a designated pressure. Modifying burners for use with a gas of unknown energy content involves compromises in efficiency and performance but should yield a usable flame. With some experimentation, you can determine the best burner characteristics for the biogas you make.

Safe Handling

When working with or around biogas (or any other combustible gas), safety is your first consideration. Methane is a highly flammable gas. It will burn when mixed with air at a ratio between 5% and 15% by volume. Biogas has a flammability range of about 4% to 25% concentration in air, and possibly wider, depending on its purity. Always take great care when working with biogas. It is unlikely, but quite possible for a biogas generator to explode, especially if there are leaks in the generator or gas piping, or if the gas pressure at the burner falls too low, allowing for flashback to occur. The entire biogas generator system is under positive pressure, and small deficiencies in the system may allow some gas to leak out. However, a catastrophic explosion is unlikely because the concentration of gas inside the system will more than likely be higher than the combustible percentage ratio. It's better to have a small amount of biogas leak to the outdoors (where it will be quickly diluted) than to have air leaking into the generator where it's more likely to mix with gas and reach the critical range of flammable ratios.

Maintain Positive Pressure

Keeping a positive pressure throughout the generator and gas lines will help to prevent air from leaking into the system, as well as prevent the burner flame from traveling through the gas line back to the digester. The normal process within the digester creates and maintains its own positive pressure as gasses are produced, but as the gas is consumed, the pressure will drop, possibly allowing flame to burn back from the burner through the delivery pipe to the generator (this is known as *flashback*). Always extinguish the burner flame if the gas pressure is too low.

Flame Arrestors

To help reduce the chances of a flashback, use a flame arrestor inside the gas pipe. The idea behind a flame arrester is to stop gas flow and/or reduce the flame temperature to below the ignition point. There are two relatively simple (though not completely foolproof) ways to approach this:

1. Incorporate a water trap into the gas line by creating U-shaped bend, or trap, in the line that is filled with a few inches of water. The gas will bubble through the water on its way out of the digester and into the gas storage container. In the event of a low-pressure situation, the water in the trap acts as physical blockage and will also cool and extinguish the flame inside the line. This approach is suitable for use between the digester and storage vessel but will not work well on the gas supply side because the flame will sputter as each gas bubble flows through the water on the way to the burner. Biogas contains water vapor that will condense in the gas line, so keep an eye on the water level in the trap.

2. This second approach works well on the gas supply line to the burner: loosely pack some fine bronze wool (similar to steel wool) into the gas supply line just before the burner. Add enough to fill 3" or 4" of line length. Gas can still move through the bronze wool, but should it start to flow back through the hose, the bronze wool

will effectively cool and extinguish the flame. Do not use steel wool as it will quickly rust, corrode, and disintegrate.

Respect the Hazards!

Treat biogas with the same respect you would treat any other fuel or flammable product. Do not work around biogas or biogas generators in the presence of sparks or flame. Never use biogas indoors or in enclosed spaces. Ethyl mercaptan is the odorant added to both natural and bottled gas so that a leak is readily detectable by smell; you do not have this safety advantage with the gas you produce. When burning biogas, ensure proper ventilation to prevent a buildup of explosive, toxic, or even deadly gases and combustion byproducts. When biogas is burned, carbon monoxide will be produced, along with CO_2, water vapor, and nitrogen oxides.

Storing Biogas

Once gas starts flowing, it can be piped into a simple holding container. For very small batch generators (of the science project variety) you can use a latex balloon. Larger amounts of gas can be stored in an open-ended barrel inverted into a larger barrel filled with water (the larger barrel on the bottom could even be a batch digester). Other storage options include tire inner tubes, UV resistant polyethylene bags, an old waterbed, or some other container of your invention. The gas storage container should be airtight, allow for expansion and contraction as gas flows into and out of it, be able to withstand and deliver the required pressure to the gas appliance, and not turn into shrapnel in the event of flashback. How much storage you need depends on how much gas you make, how much you use, and when each of those activities occurs. Greater storage capacity and isolation from the generator can be achieved using a compressor and suitable storage tank such as a portable gas grill container, but that is beyond the scope of this handbook.

Gas Burners

Compared to natural gas and propane, biogas has less energy per unit due to the dilution by CO_2 and to a much lesser extent, water vapor. This means that the energy in the flame of un-scrubbed biogas will be lower when used with conventional natural gas or LPG burning equipment. A good biogas flame requires greater fuel flow, higher pressure, and less combustion air than natural or propane gas. If you want to burn un-scrubbed biogas in conventional gas burning equipment, adjustments and modifications will be needed.

Parts of a Bunsen Burner, with the burner tube removed.

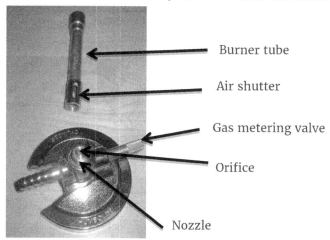

Burner tube

Air shutter

Gas metering valve

Orifice

Nozzle

Orifice Size

The burner nozzle is provided with an orifice (hole) through which the gas flows. Sometimes a nozzle can be removed and replaced with another having a different orifice. A biogas orifice will need to be larger than that used for natural or propane gas under the same pressure in order to yield a similarly energetic flame. Exactly how much larger depends on the purity of the biogas, the fuel gas you are converting from, and the pressure at which the gas will be delivered. The *Orifice Diameter Multiplier Chart* below indicates how much larger the orifice needs to be when converting equipment from natural gas or propane to biogas, assuming the same gas pressure. If you choose not to enlarge the orifice, you may still get a flame, but the energy content of the flame will be much reduced. The air-to-fuel ratio may also need to be adjusted. This can be done by experimenting with the air shutter on the burner tube, or determined by the mass flow rate of each constituent, rather than the volume ratios (as with flammability percentages).

Table 3: Orifice Diameter Multiplier for Gas Appliances

Percent Methane in Biogas	Orifice Diameter Multiplier	
	Natural Gas	Propane
70%	1.32	1.63
65%	1.39	1.72
60%	1.46	1.81
55%	1.54	1.92
50%	1.64	2.04

Here's an example: If your biogas contains 60 percent CH4, and you want to convert a natural gas appliance with an existing orifice diameter of 0.1", you need to enlarge the orifice to:

0.1 x 1.46 = 0.146" diameter

Some gas appliance nozzles can be removed and a different one installed, but many are pressed in place and cannot be removed. The nozzle orifice can be enlarged with a drill press and the appropriate drill bit.

Airflow

For a good biogas flame, the primary air supply to the burner must be reduced. This is easily accomplished by closing the air shutter on the gas appliance's burner tube until a steady, blue, cone-shaped flame is produced at the burner. The adjustment to primary air depends on the purity of the biogas.

Gas Pressure

Pressure in the biogas system can be regulated by applying external weight to the gas storage container. A pressure regulator is required on all gas appliances so that the correct gas pressure is delivered to the burner. However, finding equipment suitable for use with un-scrubbed biogas is difficult.

For reference, a typical gas kitchen range requires LPG to be delivered at 11" of water column pressure (0.4 PSI). This means that where the gas enters the appliance, it exerts the pressure required to lift a column of water in a vertical tube 11" tall. Expressed another way, 27.71 inches of water column (w.c.) is the same pressure as 1 pound per square inch (psi). Delivering biogas to a conventional cookstove burner may require up to 20" w.c. (0.72 PSI)

To give you an idea of how much weight might be required in practical terms, when my 170 lb. body compresses a volume of air under a 12"-diameter barrel inverted into a larger barrel filled with water, it exerts a pressure of about 20" w.c. After enlarging the orifice to 3/16", my Bunsen burner works well with very little gas pressure, only about 10 lbs. of weight is required on top of the inverted barrel. The flame height varies widely with pressure. A larger (20,000 Btu) burner with a ¼" orifice delivers a steady, high flame at 15" to 20" w.c. Note that gas pressure delivered to the burner is also dependent upon the diameter of the gas line.

If your gas storage system is completely sealed, it will be important to manage the system to prevent excessive pressure. A pressure relief mechanism can be as simple as an S-trap pipe filled with water that allows gas to bubble through when the pressure is too high. Relief pressure is a function of the height of the water column in the trap.

Engines

Gasoline engines can be purchased or modified for use with natural or propane gas. Modifications include installing a carburetor designed for the job, adjusting the timing, and re-gapping the spark plugs. Fuel injected engines will likely need an ignition control chip upgrade and software to access and adjust it, a topic that is beyond the scope of this discussion.

Diesel engines can generally accept methane-based gas injected via the air filter inlet so that air is mixed with up to 80% biogas.

When burning un-scrubbed biogas, the engine oil will need to be changed more frequently, and there is risk of excessive corrosion of metal engine parts. Since there is less energy in a volume of biogas compared to natural gas, gasoline, or diesel fuel, the engine will produce less power than its factory rating. Check with the engine manufacturer to ensure that there are no undue risks involved. You will likely find that any warranty will be voided if you proceed. Typically, you'll have greater success using older engines with fewer electronic controls and more user-friendly adjustments.

Purifying Biogas

Depending on the intended use and quality of the gas required, you may need to remove impurities from the gas before burning it. Biogas can contain up to 50% carbon dioxide (which won't burn), and perhaps traces of hydrogen sulfide (H_2S), which will corrode metals and break down engine oil.

Purifying, or scrubbing, biogas can decrease the detrimental effects of these ingredients, decrease gas storage requirements and increase available energy per unit of gas. There are costly high-tech gas purification systems, but our discussion will be limited to simpler methods.

One way to reduce both carbon dioxide and hydrogen sulfide in biogas is to bubble the gas through a solution of calcium hydroxide (calcium oxide and water). Calcium oxide is the chemical known as 'lime' used to fertilize soil. Hydrogen sulfide can be reduced by passing the gas through iron oxide, or rust. This can be as simple as a 4-foot long, 4-inch diameter PVC pipe filled with loosely packed rusty steel wool or coarse iron shavings. The rust will eventually be consumed and need to be replaced at a rate that is dependent upon the H_2S concentration in the biogas.

If you're making biogas for backyard grilling or tiki torch lighting, there is no need to add the complexity of scrubbing.

Environmental care

Anaerobic digestion is essentially a controlled cow fart. Both cow farts and biogas generators produce methane. Methane is a very potent greenhouse gas and should not be released directly into the atmosphere. If you are going to make methane you must burn it rather than release it into the atmosphere. Burning methane produces carbon dioxide (also a greenhouse gas but less potent than methane) and water vapour. CO_2 is also produced when organic material is composted or otherwise decomposed.

Make and Use a Biogas Generator

Making biogas usually involves smelly, messy ingredients. It's important to develop systems that work for you to minimize handling while maximizing your potential for success.

Two designs are presented here along with parts lists and assembly instructions. Please visit my <u>YouTube channel</u> (*HandsOnOffGrid*) for explanatory videos. Only basic hand tools and a power drill are required for assembly.

The first project is a very simple experimenter's batch processor made from a 5-gallon bucket. This is perfect for science experiments and for testing new recipes. I developed these kits for <u>The Vermont Energy Education Program</u> for use in their high school science curriculum. These students live in rural Vermont where fresh cow manure is plentiful!

The **hybrid system** combines elements of both the batch and continuous flow designs, allowing you to extend a single batch loading into a periodic "plug" feed so that you don't need to empty the barrel quite so often. At some point though, you will need to empty out the barrel and remove the sludge that accumulates over time. While still small in terms of gas production, I use this system at my home to process food waste and for occasional outdoor grilling.

Using plastic pails or barrels and common plumbing supply parts, both projects are fairly simple and relatively inexpensive to make and use, but they both require material processing and waste management. Don't feel limited by these designs, you can even use a plastic soft-drink bottle to start, simply adjust your expectations of gas production. The essential design elements are to provide an airtight system that allows you to put digestible material in and remove the gas produced over time.

WARNING: Making and using biogas is dangerous, potentially deadly, and can cause fires or explosions. Use common sense and proper safety equipment and procedures. Perform all experiments (including these projects) outdoors with plenty of ventilation. If you're in doubt about any procedure or have never worked with the materials and equipment described, please find an experienced helper.

For a small-scale test or science project, you can start with a simple batch digester using a 5-gallon bucket. It's a manageable project that will help you get a feel for the process. It's also useful for testing your recipe before moving on to something more complex. Follow along with the instructions and refer to the parts list and photos.

1. **Prepare the lid.**
   a. Make sure the bucket's lid gasket is in good condition for an airtight seal. If the gasket is in rough shape, remove it from inside the lid, and either replace it or clean it up, then treat it with some Petro-Gel lubrication and re-install it into the lid. This will help to ensure a good seal.
   b. Drill a hole through the top of the lid using a hole saw that's appropriately sized for the threaded bulkhead fitting. To ensure a good seal, drill the hole on a flat section of the lid without curves or raised lettering. Clean up the hole to remove any plastic shavings that could impact airtightness.
   c. Fit the threaded male end of the bulkhead fitting into the hole and secure it on the underside of the lid with the fitting's gasket and nut. Put the gasket on the flange side, not the nut side. Tighten enough to compress the gasket and ensure an airtight seal.

2. **Prepare the gas outlet and hose assembly.**

a. Wrap the threaded end of the half inch T-fitting with Teflon tape. Screw the taped end into the female threads in the top of the bulkhead fitting and tighten it snugly by hand.

b. Cut from the ½" I.D. (Inside Diameter) plastic tubing one 1" piece and two 4" pieces. These lengths can be adjusted as needed, depending on your specific requirements.

c. Fit one 4" piece of tubing onto one end of the T-fitting and secure it with a hose clamp; this is the gas supply line.

d. Attach the other end of the same tube to one side of the ball valve and secure it with a hose clamp. This valve will act as a shut off between the digester and burner.

e. Attach one end of the remaining 4"piece of tubing to the other side of the valve and secure it with a hose clamp.

f. Gently push some bronze wool into the ½" tube just after the shut off valve (downstream in the gas flow). See step 3 below for additional details.

g. The ½" gas supply line must be reduced to accept a ¼" hose to connect to the Bunsen burner's gas inlet. Install a barbed hose adapter in between the ½" and ¼" hoses and secure each connection with a hose clamp.

h. Slip the open end of the balloon through the 1" long piece of ½" tubing, then stretch the balloon over the remaining port on the T-fitting. Slide the tubing down so the end of the balloon is sandwiched between the fitting and the tube, then secure it with a hose clamp (the tube merely protects the more fragile balloon material from the clamp). The balloon serves as a gas storage vessel as well as a pressurizing system; it can also act as a safety valve by popping if the pressure inside becomes too great.

3. **Making a simple flashback preventer**. If the gas pressure drops too low, the flame at the burner can burn itself back through the gas line and into the bucket where some combustible gas remains. If the air-to-fuel ratio is right, the result can be a small fire or explosion inside the bucket. In that event, the pressure will likely pop the balloon and send up an effluent geyser. It probably won't

be catastrophic on a small-scale project like this but trust me – it's not a pretty sight.

The best way to prevent flashback is to turn off the gas supply before the pressure drops too low. "Too low" means before the flame starts to sputter as the balloon deflates.

As a precaution, create a flame arrestor by inserting some fine bronze wool (similar to steel wool) into the ½" gas supply hose. Add enough bronze wool to fill an inch or two of hose length. Don't pack it too tightly, you want gas to be able to move freely through the bronze wool. Should the gas start to flow backward through the hose, it will be cooled and extinguished by the wool.

WARNING: flashback techniques are not foolproof, so take all necessary safety precautions!

4. **Connect the Burner.**
   a. Test the system for airtightness: hold your hand over the open end of the bulkhead fitting mounted on the lid, then blow through the gas supply hose where the Bunsen burner will be connected. The balloon should inflate and there should not be any leaks.
   b. Complete the gas line connection by fitting the ¼" tubing to the gas inlet on the burner. Make sure the burner control is off.

5. **Feed and seal.** Fill the bucket halfway with ground up or finely diced food scraps, add a half-gallon or so of *very* fresh cow or pig manure, or a gallon of effluent from another operating digester as the methanogen inoculant. You could use all manure if you have enough. Add water to fill the bucket no more than ¾ full (the slurry will expand in volume as it ferments and digests) and mix it up. Securely press the lid onto the bucket and be sure it locks into place for a good seal. Test again for airtightness by blowing into the burner hose to check for leaks.

6. **Maintain the digester.** Keep the digester warm, ideally right around 100°F. The balloon will begin to inflate after a day or two,

but this first gas is mostly CO_2 and will not burn. Bleed off the gas by opening the gas valve to the burner until the balloon empties. It's best not to attempt lighting the burner at this stage because there is a greater likelihood that the air-to-fuel ratio inside the bucket is reaching the combustible zone. After a few more days and a few more balloon-fulls, the digester should be producing biogas that will burn.

7. **Light the burner.** When the balloon is full, open the gas valve on the burner and light the gas. If it sputters and doesn't light, the gas may not yet combustible, but also try different settings on the primary air inlet at the bottom of the burner. Depending on the gas pressure and the burner's orifice size, you may need to remove the burner barrel, or the orifice itself in order to observe the flame. It will be hard to see unless the light is low. If it doesn't burn, bleed off the gas from the balloon and try again tomorrow.

Keep an eye on the balloon during your burn tests, because as it deflates the pressure will drop. Remember that low pressure in the gas line can cause the flame to roll back into the bucket. Turn of the gas valve in advance!

Parts layout for the gas line

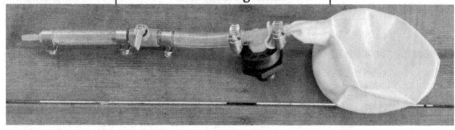

Assembled gas line

Assembled bucket

5-Gallon Parts List

Below is a list of all the parts you'll need to build the 5-gallon bucket digester with embedded Amazon product links (full disclosure – these are affiliate links and I earn a few pennies on each purchase). If you don't shop the 'zon-o-sphere, you can find many parts at local home supply stores, or online vendors such as McMaster Carr and US Plastics. At the time of this writing, all parts on this list totalled about $220, but many items are sold in package sizes with greater quantities than required.

5-Gallon Bucket Digester Parts List
5 gallon pail bucket with tight fitting lid!
Bulkhead fitting, 1/2", PVC
Mini PVC ball valve, 1/2" barb
Barbed Tube Fitting Tee for 1/2" Tube ID X 1/2 Male Pipe thread
1/2" to 1/4" tubing reducer
1/2" ID clear plastic tube (1-foot)
1/4" ID clear plastic tube (4-feet)
1/2" hose clamp (4)
1/4" hose clamp
bronze wool
balloon, sturdy, 36"
Teflon tape
Bunsen burner
insulation option 1: Reflectix
insulation option 1: black bee hive wrap
battery heat mat (optional for heat)
temperature controller (optional for heat)

Make a 55-Gallon Hybrid digester

This project uses a single wide mouth, screw-top 55-gallon (or larger) plastic barrel as the digesting vessel in what I call a hybrid biogas generator. Two additional, nested barrels are used for gas storage. The essential design elements are to provide an airtight system that lets you put digestible material in, get gas out, and allow for effluent overflow. Once you get the hang of operating this unit, you'll be able to design your own biogas generator, scaling it up or down, if desired, and adapting the system to suit your specific needs. Be sure to prepare all materials and test fit before permanently fastening.

1. **Install the effluent overflow assembly**.
   a. Drill a hole (sized for the fitting in Step 1b) through the side of the barrel, about 6" down from the top.
   b. Install a 2" unthreaded bulkhead socket fitting into the hole making sure the joint is airtight. This fitting makes the connection between 2" PVC pipes on both the inside and outside of the barrel.
      i. A flexible rubber pipe-to-tank fitting (such as those made by Uniseal) can be used as a lower cost alternative to the bulkhead adapter. Doing so will change some of the following steps.
   c. Using PVC solvent glue, install a 90-degree street elbow to the inside of the bulkhead fitting, pointing the elbow toward the bottom of the barrel.
   d. Cut a piece of 2" pipe to extend from the elbow down to about the middle of the barrel.
   e. Glue the pipe to the elbow. The bottom of this pipe must be above the bottom of the feeding tube. When the digester is filled, the pipe must be covered by the slurry at all times to prevent air from entering the barrel.
   f. Glue another 90-degree street elbow to the outside end of the bulkhead fitting, pointing down.

g. Add a pipe to the outside elbow so it extends down to an effluent collection bucket.

h. Install a clean-out with screw cap onto the end of this pipe to help keep odors down. As you load the barrel, the slurry will rise in the interior pipe until it reaches the top of the elbow and flows out, keeping the material inside the barrel at a constant level while keeping air out of the digester.

2. **Install the gas outlet.**

   a. Find a smooth, flat place in the lid and drill a hole to accept the threaded male end of a ½" threaded bulkhead fitting. Location should not interfere with the installation of the feeder pipe in Step 3 below. Install the fitting into the lid, securing it on the bottom side of the lid with the provided nut. Make sure this joint creates an airtight seal with the lid.

   b. Wrap the threaded end of the ½" x ½" NPT barb fitting with Teflon tape, and screw the fitting into the top of the bulkhead fitting.

   c. Cut a 6 to 12" piece of ½" plastic tubing, attach it to the barbed fitting and secure with a hose clamp.

   d. Connect the other end of this hose to a ball valve and secure it with a hose clamp. This valve will be used to isolate the digester from the gas storage vessel, which is important for gas pressure control. It is equally important that you remember to keep the valve open at all times *except* when the gas storage is under excess pressure. The idea is to allow gas to flow freely between digester and storage without over-pressurizing the digester. Over-pressurization may cause effluent to back up out of the feeder pipe. Not catastrophic, but certainly messy.

3. **Install the feeder tube.**

   a. In the center of the lid, cut a hole for the feeder pipe. Size it for a snug fit around the underside of a 3" toilet flange. Test fit the flange in the hole to confirm a good fit. Mark and drill four holes through the lid for mounting the flange to the lid with stainless steel machine screws.

b. Cut the feeder tube to length from 3" PVC pipe. The length of the tube depends on how tall the barrel is: the pipe will run through the toilet flange and should extend several inches above the lid at the top end (making for easier feeding and allowing for a cap to fit on top of the pipe), and down to about 12" from the bottom of the barrel when the lid assembly is installed. It's important for the bottom of the tube to be covered by the slurry inside the digester so that no air can enter.

c. Glue the PVC pipe and flange together, following the glue manufacturer's instructions, and allow to dry.

d. Apply silicone caulk liberally around the flange hole on the top side of the lid, then insert the pipe and flange through the lid and line up with the screw holes.

e. Secure the flange to the lid with four screws, washers, and nuts. Add more caulk as needed around this joint to ensure that it's airtight. Let the caulk cure completely.

4. **Insulate the barrel and add the heater (optional).** Depending on the ambient temperature surrounding the digester, you may want to provide a heat source to keep the slurry warm. For best performance, the temperature of the digester should be maintained around 100F. If you use a black barrel for the digester, it will absorb some solar heat, and a little heat will be generated by the digesting process. Housing the digester in a small greenhouse can also work well, but you may face the possibility of overheating. Take advantage of external and passive heat sources such as solar thermal, by using a heat exchange coil on the inside or outside of the digester vessel.

For my system, which lives in the colder climate of the northeastern U.S., I insulated the barrel with black poly covered foam wrap designed for insulating beehives, but you can make your own similar insulating wrap if desired. The barrel is set on top of a

2" thick piece of rigid foam insulation sandwiched between two pieces of plywood to slow heat conduction to the ground.

A submersible thermostatically controlled electric water heater is a simple way to add heat to the slurry. The heater rests inside on the bottom of the digester barrel, and the electrical cord is run up and out through the feeder pipe. A short, vertical notch cut into the top edge of the feeder pipe allows the cord to pass through so that the cap can still slide over the top of the feeder pipe.

You might find it useful to install a thermometer used for plumbing applications. A plumbing thermometer set in a "thermowell" and installed into a ½" threaded bulkhead adapter set about one-third of the way up from the bottom of the digester is suitable for this.

5. **Prepare the gas collection barrels**. The approach here is to find two barrels sized so that one can be inverted and nested into the other. The upright barrel is filled with water which acts as a seal when the other, smaller barrel, is inverted into it. All air must be evacuated from the inverted barrel so that it will fill with biogas as it is produced by the digester and received by way of the gas outlet hose and fitting installed on the bottom of the gas collection barrel. Because the barrel is inverted, the bottom faces upward. The collection barrel will rise as it fills with gas.
   a. Drill a hole in the bottom of the gas collection barrel and install ½"NPT bulkhead adapter using the same procedure as in Step 2. Wrap Teflon tape around the threaded end of the ½" NPT x hose barb T-fitting and screw it tightly into the bulkhead adapter.
   b. Fill the large barrel with water. When it's nearly full, turn the gas collection barrel upside down and push it down so that all the air escapes through the T-fitting.

6. **Make the gas line connections**.
   a. Cut a length of ½" hose to reach between the digester gas outlet and the storage barrel gas inlet. Make it long enough to create a dip in the hose that will act as a water trap.

b. Connect one end of the hose to the ball valve on the digester's gas outlet.

c. Pour a cup or two of water into the hose so that it stays at the bottom of the trap. The water in the hose will act as flashback protection for your digester should the gas in the collection barrel ignite. The clear tubing also lets you gauge the gas production rate by observing bubbles moving through the water.

d. Connect the other end of the hose to one side of the T-fitting on the gas collection barrel, securing it with a hose clamp.

e. Attach a 6 to 12" length of ½" tubing between the other end of the gas outlet T-fitting, then to a PVC ball valve. This valve controls gas flow to the gas burner and prevents it from leaking out of the storage vessel.

f. Connect a length of ½" hose to the other end of the ball valve.

g. Depending on the size requirement of the burner, a reduction adapter and a different hose diameter may be required in the gas outlet line.

h. Before making the final connection to the burner, pack some bronze wool into the gas hose to fill 1 or 2" of the hose. Pack it well, but not too tightly, so that gas can still flow through.

i. Check that all hose connections are secure, keeping in mind that you need a tight seal because the system will be under pressure.

7. **Check the system for airtightness.** Once all glues and sealants are dried and cured, completely seal the entire system as if it were in actual operation by filling with water. Pressure test using an air compressor or pump of some sort. Water in the digester should rise up the inlet tube and flow out of the effluent overflow tube, and the gas collection barrel should rise up. Operate the ball valves to isolate and test the digester and storage separately. This is an important step because you really don't want to have to deal with taking apart the system after it's loaded with manure.

8. **Load the digester.** Depending upon your location, finding a good source of inoculating manure or pond muck may be the toughest

ingredient to find. With proper management though, you may only need to do this once, as long as the methanogens stay alive. They will survive for several months without feeding.

Start by adding *fresh* cow or pig manure, or other source of anaerobic bacteria. A small is enough, a 5-gallon bucket is better, but use more if you have it. To this you can add a mixture of food scraps, grass clippings, or any combination of organic materials that result in a carbon to nitrogen ratio of somewhere between 20- and 30-to-one. Be sure the material is chopped up well for best digestion and minimum scum formation.

5 or 10 gallons of organic material slurry (including manure inoculant) is ideal to start with. The organic material will expand as it ferments and digests, and this is why the effluent overflow is not placed too close to the top of the digester barrel. It's okay to use less organic material and more water, it just means that less gas will be produced. For now, we want to create the environment in which a healthy population of methanogens can establish themselves. This can take a week or more. Observe gas production in the storage container, and check pH before feeding again.

For subsequent feedings, follow the information about recipe development and feeding, and experiment to find the best mix for your system. As a general rule of thumb, start with a 50/50 mix of water and organic material. Pour the slurry into the feeding tube through a large funnel. Use a stick as a plunger to push the material all the way down and into the barrel. When not feeding the digester, put the cap on the feeder pipe to keep odors in and air out.

9. **Test the gas for flammability**. Over the next few days to a week, biogas should start to fill the gas barrel, and it will rise up out of its water bath (be sure the gas inlet valve is open and the gas outlet valve is closed). Because there is air trapped initially inside the digester, the first barrel full of gas produced will be primarily carbon dioxide and will not burn. Simply open the gas outlet valve and release the gas from the storage barrel. Once the oxygen is

consumed and the digester turns anaerobic, combustible gases will be produced. This may take several feedings.

To safely test for flammability, sink the gas outlet hose into a bucket of water. Apply pressure to the gas barrel until bubbles rise out of the water. Have a helper hold a lighter with a long handle (such as a barbecue lighter with a long shaft) over these bubbles as they pop: if this produces a little flare you know you're producing biogas. Give it a try in the burner!

Warning! *Do not hold a match near the outlet of the digester's gas tube. With the right mixture of air and gas, a flashback could occur, and the digester could explode. You are working with a material that is as flammable, dangerous, and useful as natural gas. Use common sense, caution, and the proper safety gear.*

55-Gallon Parts List

Some of the parts required for this project such as hoses, clamps, and fittings, are the same as (or similar to) the 5-gallon bucket project. The differences are that the barrel is much larger, you'll need PVC pipes and fittings, and a suitable method of gas storage. Check local plumbing, home supply, and hardware stores for common items, and you can often find used barrels locally. Look for tanks, barrels, and fittings online at such places as US Plastic, Tank Depot, hydroponics and rain collection parts supply. McMaster-Carr is an excellent online resource for all types of hardware and bulkhead fittings (aka *through-wall pipe fittings*). Heating is highly recommended to keep the slurry temperature around 100°F. A 1,000-watt heater is sufficient to maintain the temperature of an insulated, 55-to 100-gallon digester in moderately cold climates.

If you need to buy all new parts for this project, you might spend $400–$500, depending on where you buy them. Optional equipment added to that includes a small stove to burn the biogas, and a food grinder to facilitate grinding up material for feedstock. Many of these parts are available at local plumbing supply shops and hardware stores, along with various online sources. Relatively inexpensive barrels and buckets can often be found on the local market.

55-Gallon Drum Digester Parts List

55-gallon digester barrel
water seal barrel
gas collection barrel
3" PVC pipe, 4 ft section
3" PVC cap
3" PVC toilet flange (to fit over 3" pipe)
2" PVC pipe for overflow, 5 ft section
2" PVC street elbow for overflow (2)
2" PVC cleanout with cap
2" PVC coupling
through-wall overflow option 1: 2" socket fitting for overflow
through-wall overflow option 2: 2" Uniseal
1/2" bulkhead fittings for gas outlet, inlet, and thermowell (3)
dial thermometer with stem to fit thermowell
thermowell to match thermometer and 1/2" NPT bulkhead adapter
nuts, bolts, washers to mount toilet flange to barrel top (4)
PVC glue
silicone caulk
Teflon tape
submersible heater, 1,000 watt
temperature controller (optional for heat)
food grinder, 1/2 HP
Misc. hoses, clamps, bronze wool, insulation (see 5-gallon bucket project)
Gas burner

Assembly Illustrations

Toilet flange caulked, top and bottom, then screwed to top

Feed pipe slides into flange and glued

Gas fitting through top of diester barrel

Cut slot through side of pipe for heater line

Through-wall overflow detail
Gasket goes on the non-threaded side

53

Gas fitting through top of gas collection barrel

½" barbed T fitting with ½" thread

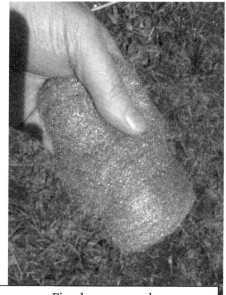

Fine bronze wool

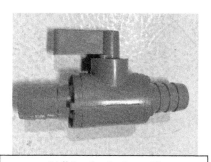

½" barbed shutoff

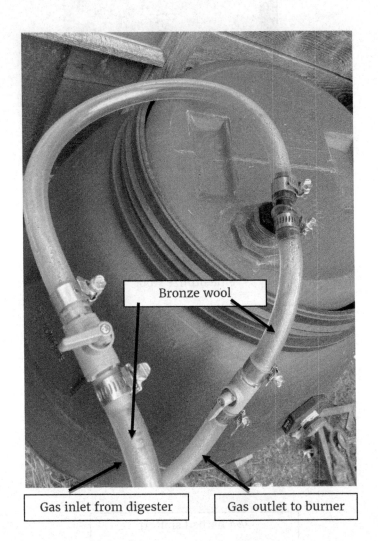

Bronze wool

Gas inlet from digester

Gas outlet to burner

Water trap in gas line

Barrel from another project, spigots not needed

Fully assembled and ready for ignition!

Food scraps and water
in, digestible slurry out

Leave room for slurry expansion and gas production! Fill to only ¾ full, max.

Beehive insulation wrapped around the digester

Gas filling the water-sealed collection barrel

Flame test!

Ready to cook with DIY Biogas!

Non-Energy Benefits

During my research into biogas, I discovered a handful of people and organizations who represent a community of dedicated clean-energy researchers and advocates. Their work goes way beyond the technical and digs deep into the issues of resource equity and how energy choices can affect the human condition. It's too easy for those of us with the convenience of automatic fuel delivery and programmable thermostats to complain about the high cost of energy, while elsewhere the cost of a hot meal ranges between unreasonable hardship and potentially life threatening.

I learned a lot of incredibly useful information from David House, author of *The complete Biogas Handbook*. In the following contribution, he discusses the non-energy benefits of renewable energy in the developing world:

A friend of mine from Sri Lanka came to visit me. He was in my office, which is packed with books, and he noticed the biogas book. He asked me about it, and I explained. He got very excited. "We need this in Sri Lanka!"

This set me on a path to researching and thinking about biogas in the developing world in a far more detailed and intensive manner than I had before. What I learned (or relearned) was that biogas is associated with strong improvements in health, education, financial well-being, gender equality, and a reduction in deforestation, among other benefits. When you think about it, that's an incredible list for something that for most of us is pretty obscure — biogas — but each item on the list has solid reasons for being mentioned:

- **Health**. The World Health Organization says that nearly two million people, mostly women and children, die every year as a result of indoor smoke from wood cooking fires. That's almost 4,500 people every day. *Biogas burns with a smokeless flame.*

- **Education**. Gathering firewood is very time-consuming (2 to 6 hours a day), and it is usually the eldest girl who does that work, so she has no time for school. Caring for a biogas digester may take a half-hour a day, and in the evening the family can have the benefit of the very bright light that biogas can provide with the right lamp, allowing for reading and study.
- **Financial well-being**. Improved health and increased time help the family (usually the mother) to start a business, barter work for an animal, or improve their financial status in other ways. Finally, the family has a crucial new increment of time and energy, both personal and household.
- **Gender equality**. The improved health and any improvements in financial strength contributed by the mother imply additional changes.
- **Deforestation**. Less wood is taken from trees cut down to feed the cooking fires. Poor countries cannot afford to replant their forests, and carbon dioxide (CO_2) released into the atmosphere from trees that are burned and not replanted has exactly the same effect as CO_2 from fossil fuels. Some studies indicate that as much as 15 percent of the increase in global CO_2 comes from those disappearing forests.

So naturally one should ask: If biogas is such a powerful tool, then why isn't it used more often? In fact, it is being used a great deal, but not nearly as much as it could be used, because the most commonly built biogas digesters are expensive: $350 to $700 for a single-family digester.

Some digester programs offer a subsidy to help purchase one, because funders and governments know what a powerful catalyst for development this technology is. Unfortunately, these programs usually leapfrog right over those who are really poor, because they want to ensure that these expensive digesters are used, and the poor do not have animals to produce manure to feed the digester. So the subsidies more often go to those who are already doing fairly well.

To address these barriers to using biogas, my work is focusing on the development of a very low-cost biogas digester. The prototypes cost about $10 in parts purchased retail in the United States in low quantity, and these digesters (like all digesters) can be fed grass, leaves, food waste, and similar materials, so they can be used by people who do not have animals.

Consider that the wealth of the world is her people, above all. We cannot doubt that people of enormous talent — nascent Mozarts, possible Einsteins, potential geniuses and savants of all stripes and kinds — have lived and died in the world's villages without the benefit of the education which would have unlocked all that was imprisoned in their hearts and minds. These gifts are lost to the whole world. In addition, poverty begets war, and one of those wars could easily suck our children into its mouth.

Addressing poverty is not merely of benefit to the world's poor, then: it will benefit everyone on this singular green planet. Imagine! If we are successful at introducing these very low-cost digesters and showing that modest profits can be made, then the idea could march across the equatorial belt like a breath of hope, helping lift families and whole villages out of poverty.

To draw attention to this project, I build and operate digester prototype digesters on my Oregon farm while building local and international community involvement with waste-to-energy management.

Cow Manure and Gas Generation; a special case

Most of us don't have a family cow at our disposal, but the following example will illustrate the process of planning a methane generator around available resources and realistic expectations. Fresh cow manure is often used exclusively for on-farm methane production and is well suited to this application. But cow manure has a less than optimal C:N of only 15:1 and relatively low gas production by total volume of raw material. Despite these unfavorable conditions, cow manure works well because it has a number of other properties that make it attractive for dairy farmers incorporating anaerobic digestion into their daily operation. These include:

- Large quantity of cow manure produced on a typical farm
- Methanogen-rich quality of cow manure, making it a guaranteed gas producer
- High water content of manure requires little additional water
- Co-benefits of capturing manure before it winds up as a pollutant in lakes and streams
- Using the digestate as a high-quality fertilizer (for on-farm use or to sell as a value-added product)
- Potential to use large amounts of gas to generate electrical power for on-farm use as well as to sell power back to the utility grid

One cow might produce about 18 gallons, (about 140 pounds) of manure each day. Given the amount of water in cow manure, only about 12% of the total weight is TS, and somewhere around 85% of TS is the useful VS portion, which may ultimately generate about 60 cubic feet of methane (or around 85 cubic feet of biogas). That amount of gas represents perhaps 3 hours of cooking fuel produced each day.

Compare the output of a dairy cow with the amount of available food scraps or human excrement (.5 lb per day per person), and the logistics of material choice for biogas production becomes clear. That's not to say you can't make useful amounts of biogas with other materials, you just need more of them. Importantly, you need to use what is available to you.

If you have one cow on your homestead property and want to collect all of its manure to make gas, you'd first need to contain the cow so you could collect the poop. If you collect 18 gallons of material, and then add enough water to make a slurry, you quickly realize that you'll need a fairly large, continuous flow biogas generator to handle a daily feeding while allowing for sufficient retention time to make a useful amount of gas.

Alternatively, a batch generator made from a 55-gallon drum loaded with a 50/50 mix of manure and water will begin producing gas within a few days, but the quantity will be limited by the amount of VS you can put in the drum.

If you've found this book helpful, please share a review so that others can join the fun of making and using biogas!

Appendix A: Characteristics of Raw Materials

Material	Type of value	%N dry weight	C:N ratio by weight	Moisture % wet weight	Bulk Density (pounds per cubic yard)	Avg %VS of TS
Crop residues and fruit/vegetable-processing wastes						
Corn cobs	Range	0.4–0.8	56–123	9–18	–	98
	Average	0.6	98	15	557	
Corn stalks	Typical	0.6–0.8	60–73	12	32	95
Fruit wastes	Range	0.9–2.6	20–49	62–88	–	75
Vegetable waste	Typical	2.5–4	11–13	62–88	–	90

Material	Type of value	%N dry weight	C:N ratio by weight	Moisture % wet weight	Bulk Density (pounds per cubic yard)	Avg %VS of TS
Manures						
Broiler litter	Range	1.6–3.9	12–15	22–46	756–1,026	Estimated average between 70 and 85% VS for most manures
	Average	2.7	14	37	864	
Cattle	Range	1.5–4.2	11–30	67–87	1,323–1,674	
	Average	2.4	19	81	1458	
Dairy tie-stall	Typical	2.7	18	79	–	
Dairy free-stall	Typical	3.7	13	83	–	
Horse-general	Range	1.4–2.3	22–50	59–79	1,215–1,620	
	Average	1.6	30	72	1379	
Horse-race track	Range	0.8–1.7	29–56	52–67	–	
	Average	1.2	41	63	–	
Laying hens	Range	4–10	3–10	62–75	1,377–1,620	
	Average	8	6	69	1479	
Sheep	Range	1.3–3.9	13–20	60–75	–	
	Average	2.7	16	69	–	
Pigs	Range	1.9–4.3	9–19	65–91	–	
	Average	3.1	14	80	–	
Turkey litter	Average	2.6	16	26	783	

Material	Type of value	%N dry weight	C:N ratio by weight	Moisture % wet weight	Bulk Density (pounds per cubic yard)	Avg %VS of TS
Municipal wastes						
Food waste	Typical	1.9-2.9	14-16	69	–	90
Night soil (humanure)	Typical	5.5-6.5	6-10	–	–	85
Paper	Typical	0.2-0.25	127-178	18-20	–	97
Refuse (mixed food, paper, and so on)	Typical	0.6-1.3	34-80	–	–	90
Sewage sludge	Range	2-6.9	5-16	72-84	1,075-1,750	87

Material	Type of value	%N dry weight	C:N ratio by weight	Moisture % wet weight	Bulk Density (pounds per cubic yard)	Avg %VS of TS
Straw, hay, silage						
Corn silage	Typical	1.2-1.4	38-43	65-68	–	most grass products average 90-95% VS
Hay-general	Range	0.7-3.6	15-32	8-10	–	
	Average	2.1	–	–	–	
Hay-legume	Range	1.8-3.6	15-19	–	–	
	Average	2.5	16	–	–	
Hay-non-legume	Range	0.7-2.5	–	–	–	
	Average	1.3	32	–	–	
Straw-general	Range	0.3-1.1	48-150	4-27	58-378	
	Average	0.7	80	12	227	
Straw-oat	Range	0.6-1.1	48-98	–	–	
	Average	0.9	60	–	–	
Straw-wheat	Range	0.3-0.5	100-150	–	–	
	Average	0.4	127	–	–	

Material	Type of value	%N dry weight	C:N ratio by weight	Moisture % wet weight	Bulk Density (pounds per cubic yard)	Avg %VS of TS
Wood and paper						
Bark-hardwoods	Range	0.10-0.41	116-436	–	–	most wood products and wood waste average 97% VS
	Average	0.241	223	–	–	
Bark-softwoods	Range	0.04-0.39	131-1,285	–	–	
	Average	0.14	496	–	–	
Corrugated cardboard	Typical	0.1	563	8	259	
Lumbermill waste	Typical	0.13	170	–	–	
Newsprint	Typical	0.06-0.14	398-852	3-8	195-242	
Paper fiber/sludge	Typical	–	250	66	1140	
Paper mill sludge	Typical	0.56	54	81	–	
Paper pulp	Typical	0.59	90	82	1403	
Sawdust	Range	0.06-0.8	200-750	19-65	350-450	
	Average	0.24	442	39	410	
Telephone books	Typical	0.7	772	6	250	
Hardwood chips, shavings	Range	0.06-0.11	451-819	–	–	
	Average	0.09	560	–	–	
Wood-softwoods (chips, shavings)	Range	0.04-0.23	212-1,313	–	–	
	Average	0.09	641	–	–	

Material	Type of value	%N dry weight	C:N ratio by weight	Moisture % wet weight	Bulk Density (pounds per cubic yard)	Avg %VS of TS
Yard wastes and other vegetation						
Grass clippings	Range	2.0-6.0	9-25	-	-	80
	Average	3.4	17	82	-	
Loose	Typical	-	-	-	300-400	
Compacted	Typical	-	-	-	500-800	
Leaves	Range	0.5-1.3	40-80	-	-	63
	Average	0.9	54	38	-	
Loose and dry	Typical	-	-	-	100-300	
Compacted, moist	Typical	-	-	-	400-500	
Seaweed	Range	1.2-3.0	5-27	-	-	50-75
	Average	1.9	17	53	-	
Tree trimmings	Typical	3.1	16	70	1296	80
Water hyacinth	Typical	-	20-30	93	405	91

Notes: Data was compiled from many references. Where several values are available, the range and average of the values found in the literature are listed. These should not be considered as the actual ranges or averages, but rather as guidance for representative values.

Table Credit: "Characteristics of Raw Materials" table taken from the *On-Farm Composting Handbook*, NRAES-54. Permission to reprint granted to the author from NRAES and the Cornell Waste Management Institute. Estimated VS content added by the author.

https://ecommons.cornell.edu/handle/1813/67142

Appendix B: Biogas Resources

The Homeowner's Energy Handbook by Paul Scheckel
  Available everywhere books are sold

Hands-On Off-Grid Author's YouTube Channel includes DIY biogas projects
  https://www.youtube.com/c/HandsOnOffGrid

The Complete Biogas Handbook by David house. If you want to become a biogas expert, get this book!
  www.completebiogas.com

Methane Recovery From Animal Manures National Renewable Energy Laboratory www.nrel.gov/docs/fy99osti/25145.pdf

The Handbook of Biogas Utilization Environmental Treatment Systems Incorporated
  www.build-a-biogas-plant.com/PDF/Handbook on Biogas
Utilization.pdf

US EPA's AgSTAR program
  www.epa.gov/agstar

Penn State Extension
  https://extension.psu.edu/agricultural-anaerobic-digesters-design-and-operation

Penn-State College of Agricultural Sciences – soil testing service
  https://agsci.psu.edu/aasl/compost-testing
Search for agricultural testing service near you

3-cubic meter biogas plant Volunteers in Technical Assistance (now defunct) Technical Publication
www.builditsolar.com/Projects/BioFuel/VITABIOGAS3M.HTM

Appendix C: Quick Unit Conversions

100°F = 37.8°C
20°C = 68°F
1-cubic foot = 1,728 cubic inches
1-cubic foot = 0.028 cubic meters
1-cubic meter = 35.31 cubic feet
1-pound = 0.45 kilograms
1-kilogram = 2.2 pounds
1-inch = 2.54 centimeter
1-BTU = 1,055 Joule

Notes

Made in the USA
Middletown, DE
15 June 2023

32651879R00046